D0893861

Spearville
Grade School
Library

A PICTURE BOOK OF
COWS

Spearville
Grade School
Library

by Dorothy Hinshaw Patent
photographs by William Munoz

HOLIDAY HOUSE / NEW YORK

This book is lovingly dedicated to PEGGY, JACK, and
SANDY, without whom the photographer would not be.

ACKNOWLEDGMENTS

The author and photographer would like to thank the following people for generously
giving their valuable time to share their knowledge of cows and cattle: Betty Jo Johnson,
Pat Holt, Buddy and Cindy Westphal, John Muller, Mrs. Roy VanOstrand, Jim and Julie
Cusker, Sjaan Vincent, and Paul Guenzler and his family. The following people were kind
enough to let us use photos of them and/or their animals: Roger and Kay Bowers (pp. 4, 5,
6, 7, 8); Dale and Betty Jo Johnson, Dale's Dairy (pp. 9, 15, 18, 20, 21, 25, 26); Jane Spahr
(pp. 9, 28, 39); The William Holt family (pp. 10, 31); Roy Tufly (p. 12); Buddy and Cindy
Westphal (pp. 13, 30); John Muller (pp. 14, 29); Roy VanOstrand (pp. 14, 20); Jack Munoz
and Sandy Munoz (p. 14); Western Montana Production Credit Association (p. 18); Jim
Cusker and Julie Cusker (p. 21); The Gary Simonson family (Tammy, p. 22; p. 23; Gina,
p. 24); Dave and Sjaan Vincent (pp. 27, 35, 36); Paul Guenzler and Russ Sherman (p. 32);
Mort Lytle (p. 33); and Forrest Davis (pp. 34, 38).

Text copyright © 1982 by Dorothy Hinshaw Patent
Photographs copyright © 1982 by William Munoz
All rights reserved
Printed in the United States of America
First Edition

Library of Congress Cataloging in Publication Data

Patent, Dorothy Hinshaw.
    A picture book of cows.

    Includes index.
    Summary: Examines the differences between dairy cows
and beef cattle; identifies the most popular breeds;
and covers such topics as raising calves, milking,
and branding.
    1. Cows—Juvenile literature. 2. Cattle—Juvenile literature.
[1. Cows.   2. Cattle] I. Munoz, William, ill. II. Title.
SF197.5.P37   636.2   82-80819
ISBN 0-8234-0461-7   AACR2

81-476   (1984-85)

Contents

Being Born

Most calves are born during the chilly months of late winter or early spring. The mother cow lies on her side. Her baby is born with a thin sac covering its body. The calf has been growing inside its mother for nine months.

After her calf is born, the new mother stands up. She turns around and licks her baby. She cleans off the sac and fluffs up the calf's fur as she dries it with her tongue.

Only minutes after it is born, the wet little calf tries to stand up. It struggles to put its long shaky legs under its body. Standing up isn't easy when you are only fifteen minutes old!

But soon the calf is on its feet. Just standing is hard work. The cow moos hopefully to her baby. The calf hears her voice and will remember just how it sounds. Each calf learns its own mother's special "moo." Then it can find her even in a large herd.

Growing Up

Most cows are good mothers. They stay close to their young calves and protect them from strangers. They lick them to keep them clean. Many ranchers put tags on the ears of their cows and calves. Then it is easy to tell the animals apart.

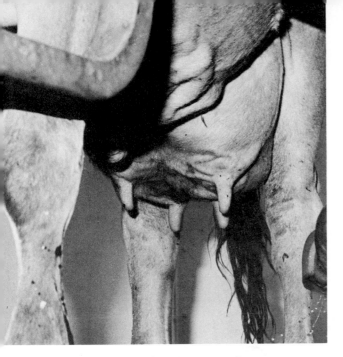

Each mother cow has an udder that makes milk to feed her calf. The udder has four teats shaped like fingers. The calf sucks milk through the teats. The mother cow lets her calf nurse whenever it wants.

The udder has four teats from which the calf can suck milk.

As the calves get older, they spend less time with their mothers. They make friends with other calves. They learn to eat grass and hay.

Many farm and ranch children raise calves. The children work hard taking good care of their animals. In late summer, the children take their calves to the county fair to be judged. They wash them and brush them until their coats shine. Then they lead them before the judges. The judges decide which calves they think are best.

Being a Cow

When ranchers talk about their cows, they call them cattle. The word *cattle* includes males and females, young and old. The word *cow* is used mostly for grown-up female cattle. Adult male cattle are called bulls. Bulls can breed with cows to produce calves. About half of the calves that are born are bull calves. But only one bull is needed to breed with many cows. As bull calves grow up, they fight with each other and are hard to keep together. So ranchers turn most of their bull calves into steers.

A bull

A heifer

Steers are males which have been "castrated" (KAS-tray-ted). They have had their male sex glands, the testes (TEST-eez), removed. Steers are gentler than bulls and can be raised together in pastures and pens, without fighting. Steers cannot mate with cows to produce calves. They are raised by the ranchers for their meat.

Young cows are called heifers (HEF-erz). Ranchers keep most of their heifers. They let them grow up into cows so they can have calves.

Cattle are big animals. Adults can weigh between 1,000 and 3,000 pounds.

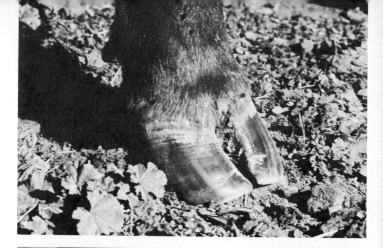

Their hooves have two big toes which support their heavy bodies.

Their damp noses sniff at food and smell other animals. A cow can recognize her calf by how it smells.

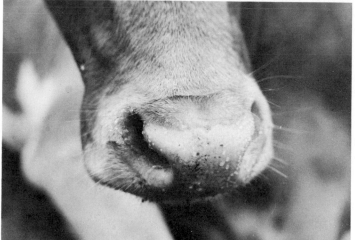

Their long tongues pull up clumps of grass. Like all animals, cattle enjoy treats such as fresh apples.

Their long tails swish away the flies that lay eggs under their skin or bite them.

One reason cattle have such big bodies is because of their huge stomachs. Their stomachs are divided into four parts.

The first two parts form a large pouch called the rumen (ROO-min). When the cow swallows, its food passes into the rumen. After the cow has eaten a lot, it coughs up some food from the rumen. It chews it with its strong back teeth to make it easier to digest. Then the cow swallows the food again, bringing up another batch for chewing. This is called "chewing its cud."

Now and then, food from the rumen passes into the third stomach part. Here it is digested some more. Finally, the food is carried to the last part of the stomach, where the cow finishes digesting it.

Cattle spend a third of their time resting, a third of their time eating, and a third of their time chewing their cuds.

Cows often lie down as they chew their cuds.

Oxen are pulling this wagon

Cattle are very useful animals. In many countries they work on farms, pulling plows, carts, and wagons. Working cattle are usually steers and are called oxen. Cattle hides make nice leather, and cow manure is good for the soil. But the most important things about cattle are their milk and their meat. Some cattle are bred for their milk; others for their meat. Cattle live very different lives, depending on if they are dairy cows or beef cattle.

Dairy Cows

Many breeds of cattle are raised as milk or dairy cows. Energy from their food goes into making milk instead of body fat. That's why they have lean, bony bodies. The most popular dairy cow is the Holstein (HOLE-steen). It makes more milk than any other breed. Holsteins are large and have black and white blotches on their bodies. A few Holsteins are red and white.

Holstein milk is pale and does not have much cream in it. So dairies often have other kinds of cows which produce creamier milk.

Holstein

Guernseys (GERN-zees) make rich milk with a fine golden color. Good yellow butter can be made from it. They are gentle animals with tan and white bodies. You can see the bony look of a good dairy cow in this Guernsey.

Ayrshire (AIR-shir) cows make rich milk. They usually have red and white markings.

Brown Swiss cattle have brown bodies with black hooves. They have white muzzles, dark noses, and big ears. Brown Swiss make very white milk and are easy to care for.

Jerseys are good family cows, since they are small and give rich, creamy milk. This Jersey has a fuzzy winter coat.

Dairy calves are taken away from their mothers when they are only a day or two old. Then they are fed milk from a bottle for two or three months. Dairy calves are also given hay and grain when they are about a week old.

Spearville
Grade School
Library

These Holsteins are enjoying their warm, steamy food.

Dairy cows are fed plenty of food. The more they eat, the more milk they make. One cow can eat 30 pounds of hay and 20 pounds of grain in one day. This is a lot, but the cow will give the farmer 50 pounds or more of milk in return. That is almost six gallons.

23

Dairy cows are milked twice a day. When milking time comes, the cows crowd together. They are brought a few at a time into the "milking parlor." First their legs and udders are washed. Then the very first drops of milk are squeezed out by hand.

A milking machine is attached to the teats. It massages them, pulling the warm milk into hoses which lead to cooled storage tanks.

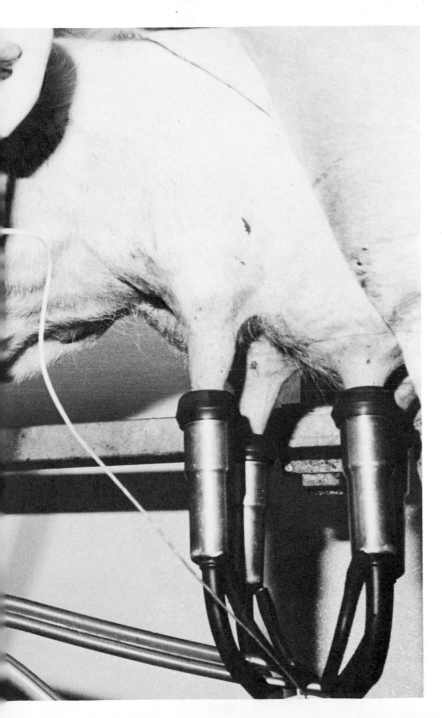

The milking machine pumps milk from the teats into hoses which lead to a cooling tank.

A machine quickly fills the bottles with milk and caps them.

When the milk is ready to be put into bottles, it is pasteurized (PAS-chur-ized). It is heated briefly to 170 degrees so that any germs are killed. It is also shaken up to make the cream mix with the milk. This step is called homogenization (ho-moj-eh-nih-ZA-shen). Then the milk is pumped into bottles or cartons which are taken to stores and sold.

Beef Cattle

Beef is the most popular meat in the United States. Steaks and hamburgers are enjoyed by many people and come from the muscles of beef cattle. When beef cattle feed, the energy from their food is used to make them grow fast and put on muscle and some fat. Beef cattle have long, straight backs and sturdy bodies. Their legs are usually shorter than those of dairy cows.

These beef cattle are called Herefords (HER-furds). They have a thick, curly, red-brown coat, white faces, and other white markings. Herefords are popular on ranges in the West. They are calm and grow well on a diet of grass.

Angus (AN-gus) cattle thrive on the rugged western range. Most Angus are solid black, but there is also a breed called Red Angus. Angus cows have small calves and are good mothers.

Charolais (shar-o-LAY) cattle from France are becoming more and more popular with ranchers. These gentle, pretty white cattle make lots of milk for their calves. Charolais calves are bigger and grow faster than Angus or Herefords. They are ready to sell for beef at a much earlier age.

During the 1800s, Longhorn cattle were raised in Texas and driven to Kansas to be sold. When other breeds became popular, Longhorns became rare. But now the breed is growing again. Longhorns come in many different colors and have beautiful long, curved horns.

A Simmenthal bull

Several large breeds of beef cattle from Europe such as the Simmenthal (ZIM-men-tal) and Limousin (Lim-eh-ZEN) are being bred in the United States. The calves grow up quickly, becoming large animals with lots of meat. They are good for ranchers who want to raise as much meat as possible.

Brahmans are large and have loose skin and big, floppy ears. The bulls have a hump on their shoulders. Brahmans can be bred to other cattle to create breeds such as the Brangas (Brahman plus Angus). The Santa Gertrudis (gur-TROO-dis) breed was developed by crossing Brahman and Shorthorn cattle.

This Brahman bull is much bigger than the Hereford bull standing in front of him.

This is a beefalo calf. It is the result of crossing cattle with bison (BY-sun), which are also called American buffalo. Beefalo are very hardy and grow quickly. They do better on poor feed than regular cattle and can survive well in cold or hot climates. Beefalo meat is leaner than beef.

The life of a beef calf is different from that of a dairy calf. The beef calf stays with its mother after it is born. When the beef calves are two or three months old, they are brought in for branding. First the cows and calves are rounded up and separated from one another. The calves peer through the fence, looking for their mothers. The cows moo and moo while they wait for their babies. The air is filled with the sounds of cattle.

One by one the calves are clamped onto the branding table. Each ranch has its own special brand. A hot iron is used to burn a mark into the skin of the frightened calf. Then everyone will know who owns it if it gets through a fence and is lost. At the same time, the calf is given a shot to protect it from diseases. The male calves are castrated to turn them into steers. The branding hurts, but it only takes a few seconds. Then the calf is freed to go find its mother again.

A calf is given a shot as it is branded.

After branding is over, the cattle are left out on the range to feed and grow. When the calves are about six months old, they are separated from their mothers again. The mothers moo for their babies, and the calves cry for their mothers. But they are now ready to be on their own. The mother cows' milk dries up and the calves become used to eating only hay and grass. After about a month, the calves and cows can be put back together again.

During the summer and fall, most beef cattle are left to feed on the range. Later, they are moved to pens, where they are fed grain to fatten them up. The steers are sold for meat when they reach about 1,000 pounds.

When winter comes and the grass stops growing, ranchers bring their cattle into pastures near the ranch buildings. Every morning until springtime, they must feed the cattle.

Horses are often used to pull cattle-feed wagons.

In late winter or early spring, calving time arrives again. A new batch of tiny calves is born. Each calf struggles to its feet to greet the world, and a new year for the cattle has begun.

Index